~A BINGO BOOK~

Chemistry Bingo Book

COMPLETE BINGO GAME IN A BOOK

Written By Rebecca Stark

ISBN 978-0-87386-447-3

Educational Books 'n' Bingo

Printed in the U.S.A.

CHEMISTRY BINGO DIRECTIONS

INCLUDED:

List of Terms

Templates for Additional Terms and Clues

2 Clues per Term

30 Unique Bingo Cards

Markers

1. **Either cut apart the book or make copies of ALL the sheets. You might want to make an extra copy of the clue sheets to use for introduction and review. Keep the sheets in an envelope for easy reuse.**

2. Cut apart the call cards with terms and clues.

3. Pass out one bingo card per student. There are enough for a class of 30.

4. Pass out markers. You may cut apart the markers included in this book or use any other small items of your choice.

5. Decide whether or not you will require the entire card to be filled. Requiring the entire card to be filled provides a better review. However, if you have a short time to fill, you may prefer to have them do the just the border or some other format. Tell the class before you begin what is required.

6. There are 50 terms. Read the list before you begin. If there are any terms that have not been covered in class, you may want to read to the students the term and clues before you begin.

7. There is a blank space in the middle of each card. You can instruct the students to use it as a free space or you can write in answers to cover terms not included. Of course, in this case you would create your own clues. (Templates provided.)

8. Shuffle the cards and place them in a pile. Two or three clues are provided for each term. If you plan to play the game with the same group more than once, you might want to choose a different clue for each game. If not, you may choose to use more than one clue.

9. Be sure to keep the cards you have used for the present game in a separate pile. When a student calls, "Bingo," he or she will have to verify that the correct answers are on his or her card AND that the markers were placed in response to the proper questions. Pull out the cards that are on the student's card keeping them in the order they were used in the game. Read each clue as it was given and ask the student to identify the correct answer from his or her card.

10. If the student has the correct answers on the card AND has shown that they were marked in response to the *correct questions,* then that student is the winner and the game is over. If the student does not have the correct answers on the card OR he or she marked the answers in response to *the wrong questions,* then the game continues until there is a proper winner.

11. If you want to play again, reshuffle the cards and begin again.

Have fun!

TERMS

ABSORPTION

ACIDS

ATOM

ATOMIC NUMBER

BASES

BOND

BOYLE'S LAW

BUNSEN BURNER

CARBON

CATALYST

CHEMISTRY

CHLORINE

COMPOUND

CORROSION

DENSITY

DISTILLATION

ELECTRICAL CHARGE

ELEMENTS

ENERGY

EQUILIBRIUM

GAS

HELIUM

HYDROGEN

ION

LIQUID

MASS

MATTER

DMITRI MENDELEEV

METALS

MIXTURE

MOLE

MOLECULES

NITROGEN

OSMOSIS

OXIDATION

OXYGEN

PERIODIC TABLE

pH

POLYMER

PROPERTIES

REACTANTS

REACTION

SATURATION

SCIENTIFIC METHOD

SODIUM

SOLID

SOLUTION

SURFACE TENSION

TEMPERATURE

VACUUM

Additional Terms

Choose as many additional terms as you would like and write them in the squares. Repeat each as desired.
Cut out the squares and randomly distribute them to the class.
Instruct the students to place their square on the center space of their card.

Chemistry Bingo

Clues for Additional Terms

Write three clues for each of your additional terms.

1. _____ 2. 3.	1. _____ 2. 3.
1. _____ 2. 3.	1. _____ 2. 3.
1. _____ 2. 3.	1. _____ 2. 3.

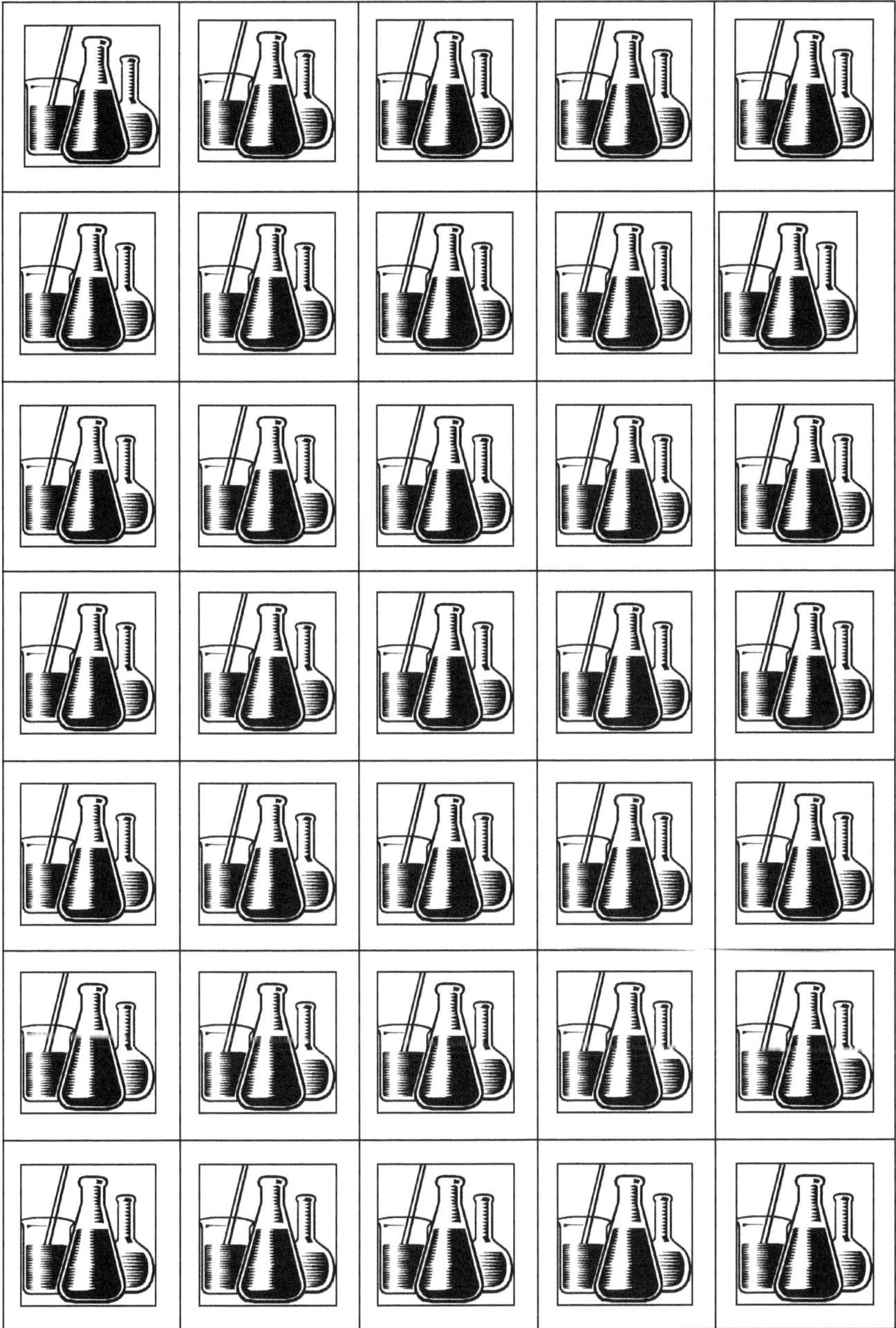

ABSORPTION 1. It is the penetration of molecules into the bulk of a solid or liquid, forming either a solution or compound. 2. It differs from adsorption because in adsorption molecules are not penetrated into the interior. 3. This is a process in which one substance permeates another.	**ACIDS** 1. They have a sour taste. Examples are lemon juice and vinegar. 2. They turn litmus paper red. 3. These substances can release a hydrogen ion.
ATOM 1. It is the smallest component of an element. 2. Protons, which carry a positive charge, and neutrons, which carry no charge, make up its center, or nucleus. 3. Electrons, which have a positive charge, circle its nucleus.	**ATOMIC NUMBER** 1. This is the number of protons in the nucleus of an atom. 2. The Periodic Table of Elements is arranged according to this. 3. It determines which element an atom is.
BASES 1. These substances can accept a hydrogen ion from another substance. 2. They turn litmus paper blue. 3. Strong ones, such as oven cleaners, are very dangerous if swallowed.	**BOND** 1. This is the result when forces of attraction hold together atoms in an element or compound. 2. Ionic, covalent, metallic, and intermolecular are types of this. 3. It is the attractive force that holds together ions, atoms or groups of atoms.
BOYLE'S LAW 1. This law states that gas compresses proportionately to the amount of pressure exerted on it. 2. According to this law, if you have a 1-cubic-foot balloon and double the pressure on it, it will be compressed to 1/2 cubic foot. 3. This gas law was named after the scientist who originally published it in 1662.	**BUNSEN BURNER** 1. This common laboratory equipment uses natural gas or propane to produce an intensely hot flame. 2. Air enters it through adjustable holes and produces a very hot blue flame. 3. This laboratory equipment was named for a nineteenth-century German chemist.
CARBON 1. This element is found in all known life forms. 2. This nonmetal element has the symbol C; its atomic number is 6. 3. Three forms of this element are diamond, graphite, and coal.	**CATALYST** 1. It is a substance that alters the speed of a chemical reaction but is left unchanged. 2. This substance does not appear in the final product of a chemical reaction and undergoes no permanent changes. 3. The opposite of this is an inhibitor, which slows the rate of a reaction without being consumed.

Chemistry Bingo

CHEMISTRY
1. This branch of science deals with the composition of substances and their properties and reactions.
2. The study acids and bases is included in this branch of science.
3. The Periodic Table is used in this branch of science.

CHLORINE
1. This element is important to digestion because stomach acid, or HCl, is a compound of hydrogen and this.
2. Along with sodium and potassium, this element is called an electrolyte because it carries an electrical charge when dissolved in body fluids.
3. It is a halogen, or salt-former. Table salt, NaCl, is a compound of sodium and this element.

COMPOUND
1. This is a distinct substance formed by the chemical union of two or more ingredients.
2. Water is one. It is made up of two hydrogen atoms bonded to one oxygen atom.
3. Carbon dioxide is a chemical one; it is composed of two oxygen atoms bonded to one carbon atom.

CORROSION
1. This is a reaction between a material, usually a metal, and its environment that results in a deterioration of the material.
2. The most common form of this is rusting, which occurs as a result of the oxidation of iron.
3. This is the wearing away of metal by a chemical reaction.

DENSITY
1. This is a measure of the amount of mass per unit of volume. The formula to find it is *mass/volume.*
2. Mass is a measure of how *much* matter there is in an object; this is a measure of *how tightly* that matter is packed together.
3. An object will sink if its ___ is greater than the fluid; it will float if its ___ is less than the fluid.

DISTILLATION
1. This process purifies a liquid by successive evaporation and condensation.
2. In this process a mixture is heated to separate the more volatile parts from the less volatile. Then the resulting vapor is cooled and condensed.
3. This process is used to refine crude oil. The different hydrocarbon chains are separated by their vaporization temperatures for different uses.

ELECTRICAL CHARGE
1. A fundamental property of matter, this may be positive, negative or zero.
2. The energy made available by its flow through a conductor is called electricity.
3. When an atom has the same number of protons and electrons, this is zero and the atom is said to be neutral.

ELEMENTS
1. These are substances whose atoms are all the same. Their symbols usually consist of the first one or two distinctive letters in the name.
2. These cannot be decomposed into two or more other ones by means of chemical change.
3. Most ___ are metals. Examples of metallic ones are gold, silver, copper, & mercury. An example of a nonmetallic one is oxygen.

ENERGY
1. In the scientific sense, it may be thought of as the capacity to change the motion of objects.
2. It can change form and can be transferred from one body or system to another, but its total amount remains the same.
3. Its many forms include potential, or stored; kinetic, or movement; chemical; heat; light; electrical; sound; and nuclear.

Chemistry Bingo

EQUILIBRIUM
1. This has been reached in a chemical reaction when the rate of the forward reaction is equal to the rate of the reverse reaction.
2. When a chemical reaction has reached this, collisions are still occurring but the reaction is now happening in each direction at the same rate.
3. If the rate of evaporation equals the rate of condensation, this relationship exists.

GAS 1. In this state, or phase, matter takes the size and shape of its container. 2. In this state, particles in the matter have no definite shape and move around freely. 3. In this state, matter expands and contracts with changes in temperature or pressure; therefore, this state is said to be compressible.	**HELIUM** 1. The symbol of this inert gas is He. 2. This element is used to inflate lighter-than-air balloons. Although heavier than hydrogen, this gas is safer because it does not burn. 3. Along with neon, argon, krypton, xenon and radon, it is a member of Group 18 of the Periodic Chart. Group 18 comprises the non-metal inert elements called Noble Gases.
HYDROGEN 1. It is the lightest element. It is by far the most abundant element in the universe. 2. It combines with oxygen to make H_2O, or water; therefore, it is essential to life. 3. It has the atomic number 1.	**ION** 1. It is an atom or a group of atoms with a net electric charge because of gaining or losing one or more electrons. 2. It is an atom with an unequal number of protons and electrons. 3. The process by which a neutral atom or a cluster of neutral atoms becomes one is called ionization.
LIQUID 1. In this state, particles are close together with no regular arrangement. 2. In this state, particles can move and slide past each other. The matter assumes the shape of the part of the container it occupies. 3. Matter in this state, like in the solid state, has little free space between particles and is not easily compressible.	**MASS** 1. This is the property that causes matter to have weight in a gravitational field. On Earth's surface it is the same as weight. 2. It is the amount of matter in something. 3. The SI unit of this is the kilogram (kg).
MATTER 1. It occurs in three main states, or phases: solid, liquid and gas. 2. Its fourth state, plasma, occurs when electrons are stripped away by high heat or pressure. 3. It changes from a solid state to a liquid state at its melting point.	**DMITRI MENDELEEV** 1. Although others made contributions, this Russian chemist is known as the "Father of the Periodic Table." 2. He published his periodic table in *Principles of Chemistry* in 1869. 3. Not only did he leave spaces for elements not yet discovered, he also predicted properties of 5 of these elements and their compounds.
METALS 1. About 75% of the known elements are these. 2. They are malleable; in other words, they can be stretched, drawn or hammered; they are good thermal and electrical conductors; and they have luster. 3. Examples are lithium, in the Alkali Group; magnesium, in the Alkaline Earth Group; and gold, in the Transition Group. Chemistry Bingo	**MIXTURE** 1. It consists of two or more pure substances that are combined but not bonded. A solution is one type. 2. Each component of this retains its own chemical properties and may be present in any proportion. 3. A solution is a homogeneous one. A suspension is a heterogeneous one. © **Barbara M. Peller**

MOLE 1. It is the name for a quantity of particles of any type equal to Avogadro's number: 6.02×10^{23} particles. 2. This term is often used in place of "gram-molecular weight." 3. This unit of measure is used to compare quantities of different substances. It is a unit in the International System of Units, or SI.	**MOLECULE** 1. It is the simplest structural unit of an element or compound and is made up of even smaller particles called atoms. 2. It is a unit of atoms bonded together. 3. Aspirin is a common one. It is made up of 9 carbon atoms, 8 hydrogen atoms and 4 oxygen atoms.
NITROGEN 1. This chemical element has the symbol N and the atomic number 7. 2. This colorless, odorless, and generally inert gas makes up about 78% of Earth's atmosphere by volume. 3. Although Lavoisier named it *azote,* meaning "without life," its compounds are vital components of foods, fertilizers, & explosives.	**OSMOSIS** 1. It is the diffusion of water or other solvent across a semi-permeable membrane. 2. During this process the solvent moves from a region of low concentration of solute to one with a high solute concentration. 3. In this process solvent molecules move from a dilute solution into a more concentrated onen, causing the latter to become more dilute.
OXIDATION 1. This process is the interaction between oxygen molecules and the substances they contact. 2. This process can cause a freshly cut apple to turn brown. 3. This process can result in corrosion.	**OXYGEN** 1. This colorless, odorless, and tasteless gas has the atomic number 8 and is represented by the symbol O. 2. The 3rd most abundant element in the universe, it makes up about 21% of our atmosphere. 3. Joseph Priestley and Carl Wilhelm Scheele both independently discovered this element, but Priestley is usually given credit for the discovery.
PERIODIC TABLE 1. Mendeleev is credited with the invention of this tabular method of displaying chemical elements. 2. This table arranges chemical elements according to their atomic numbers. 3. It groups elements with similar properties, such as Alkali Metals, Alkaline Earth Metals, Transition Metals, Metalloids, Non-Metals, Halogens, Noble Gases and Rare Earth Elements.	**pH** 1. This scale is used to measure the acidity or alkalinity of a solution. 2. The letters stand for "potential of hydrogen." 3. This scale ranges from 0 to 14, where 7 is considered neutral. Pure water is 7. Below 7 is acidic and above 7 is basic.
POLYMER 1. This is one of many natural and synthetic compounds consisting of repeated structural units that are linked by chemical bond. 2. This large molecule is created by many smaller molecules, called monomers, in a regular pattern. 3. Examples include plastics, DNA and proteins. Chemistry Bingo	**PROPERTIES** 1. These are the characteristics of a substance. 2. They are things you can observe, such as size, color, shape, or texture. 3. If we say that gold is malleable, ductile, and soft and that it is a good conductor of heat and electricity, we are describing these.

REACTANTS 1. These are substances that participate in a chemical reaction. 2. Sulfur and iron are these in the chemical reaction resulting in the formation of iron sulfide. 3. In the chemical reaction known as corrosion, iron and oxygen are these. They combine to form iron oxide, or rust.	**REACTION** 1. One that involves oxidation and reduction is called redox. 2. New substances formed as a result of one are called products. 3. If energy in the form of heat, light, or sound is released, it is called an exothermic one. If it must absorb energy in order to proceed, it is called an endothermic one.
SATURATION 1. This is the point at which a solution of a substance can dissolve no more of that substance. 2. At this point additional amounts of a substance added to a solution will appear as a precipitate. 3. When referring to this in regards to the nitrogen in an ecosystem, it means that the soil cannot store any more nitrogen.	**SCIENTIFIC METHOD** 1. This process is the basis for scientific inquiry. 2. This process follows a series of steps: identify a problem, formulate a hypothesis, test the hypothesis, collect and analyze the data, and make conclusions. 3. This procedure for analyzing scientific problems in a way that leads to verifiable results is based upon controlled experiments.
SODIUM 1. Its symbol is Na; its atomic number is 11. 2. This element is in Group 1, often referred to as the "Alkali Metals." Other members of Group 1 include lithium, potassium, rubidium, caesium, and francium. It also includes hydrogen; however, hydrogen rarely acts like an alkali metal. 3. Its most common form is NaCl, or table salt.	**SOLID** 1. In this state, or phase, matter retains a fixed volume and shape. 2. Like the liquid state, this state is often referred to as a condensed phase because the particles are very close together. 3. A crystal is one in which which the atoms or molecules are arranged in a definite pattern that is repeated regularly in three dimensions.
SOLUTION 1. This is a mixture formed when a material dissolves in a liquid and cannot be filtered out. 2. The component present in the greatest amount in one of these is called the solvent. 3. The substance that dissolves in a liquid to form one is called the solute.	**SURFACE TENSION** 1. This force draws the surface molecules into the bulk of a liquid causing the liquid to assume the shape having the least surface area. 2. It is a property of the surface of a liquid that causes it to behave like an elastic sheet. 3. It is the increased attraction of molecules at the surface of a liquid.
TEMPERATURE 1. As used in chemistry and other physical sciences, this is a measurement of the average kinetic energy in a sample—in other words, how fast the molecules are vibrating. 2. Three scales used to measure this value are Kelvin, Celsius, and Fahrenheit. 3. We commonly refer to it as the degree of hotness or coldness of a body or environment. Chemistry Bingo	**VACUUM** 1. It is a space essentially empty of matter. Pressure in one is less than atmospheric pressure. 2. In reality, no volume of space can ever be a perfect one because there cannot be a gaseous pressure of absolute zero. 3. It is the absence of air or other gas.

Chemistry Bingo

Electrical Charge	Absorption	Bond	Ion	Dmitri Mendeleev
Boyle's Law	Acids	Solid	Mole	Surface Tension
Atom	Vacuum		Osmosis	Chemistry
Reaction	Distillation	Temperature	Mass	Nitrogen
Oxygen	Equilibrium	Periodic Table	Solution	Scientific Method

Chemistry Bingo: Card No. 1

Chemistry Bingo

Reaction	Atom	Mixture	Saturation	Hydrogen
Nitrogen	Compound	Atomic Number	Distillation	Matter
Chlorine	Equilibrium		Elements	Temperature
Polymer	Oxidation	Vacuum	Properties	Scientific Method
Surface Tension	Solid	Periodic Table	Boyle's Law	Solution

Chemistry Bingo: Card No. 2

Chemistry Bingo

Reaction	Temperature	Compound	Mass	Atom
Mole	Acids	Catalyst	Absorption	Liquid
Distillation	Solid		Matter	Bases
Vacuum	Chlorine	Oxygen	Polymer	Mixture
Solution	Boyle's Law	Periodic Table	Properties	Hydrogen

© Barbara M. Peller

Chemistry Bingo

Vacuum	Matter	Bond	Boyle's Law	Hydrogen
Metals	Carbon	Absorption	Saturation	Atom
Osmosis	Polymer		Dmitri Mendeleev	Ion
Temperature	Corrosion	Solid	Periodic Table	Atomic Number
Bunsen Burner	Surface Tension	Oxidation	Solution	Chemistry

Chemistry Bingo

Surface Tension	Dmitri Mendeleev	Distillation	Atomic Number	Boyle's Law
Metals	Temperature	Catalyst	Elements	Acids
Bond	Chemistry		Mole	Gas
Scientific Method	Hydrogen	Electrical Charge	Properties	Density
Compound	Periodic Table	Atom	Vacuum	Osmosis

Chemistry Bingo

Bases	Matter	Mixture	Hydrogen	Chemistry
Mass	Distillation	Density	Absorption	Atom
Saturation	Bunsen Burner		Carbon	Elements
Periodic Table	Oxygen	Properties	Oxidation	Bond
Nitrogen	Atomic Number	Electrical Charge	Osmosis	Corrosion

Chemistry Bingo

Electrical Charge	Matter	Gas	Mole	Compound
Nitrogen	Hydrogen	Equilibrium	Acids	Metals
Mixture	Ion		Elements	Carbon
Vacuum	Polymer	Catalyst	Reaction	Chlorine
Periodic Table	Boyle's Law	Properties	Oxidation	Bases

Chemistry Bingo: Card No. 7

Chemistry Bingo

Osmosis	Matter	Energy	Mass	Carbon
Metals	Bond	Saturation	Chemistry	Atomic Number
Corrosion	pH		Hydrogen	Dmitri Mendeleev
Solution	Vacuum	Reaction	Bunsen Burner	Polymer
Solid	Periodic Table	Oxidation	Distillation	Nitrogen

Chemistry Bingo: Card No. 8

Chemistry Bingo

Elements	Compound	Equilibrium	Corrosion	Boyle's Law
Bunsen Burner	Hydrogen	Osmosis	Distillation	Matter
Liquid	Electrical Charge		Acids	Energy
Density	Scientific Method	Oxygen	Mole	Gas
Polymer	Properties	Catalyst	Reaction	Dmitri Mendeleev

Chemistry Bingo

Reaction	Mass	Carbon	Saturation	Corrosion
Chemistry	Atomic Number	Absorption	Acids	Hydrogen
pH	Matter		Ion	Chlorine
Oxygen	Scientific Method	Density	Properties	Liquid
Catalyst	Nitrogen	Mixture	Surface Tension	Osmosis

Chemistry Bingo: Card No. 10

Chemistry Bingo

Bases	Matter	Distillation	Density	Nitrogen
Energy	Liquid	Mole	Elements	Absorption
Metals	Hydrogen		Mixture	Equilibrium
Catalyst	Atom	Properties	Boyle's Law	Reaction
Bunsen Burner	Periodic Table	Electrical Charge	Oxidation	Compound

Chemistry Bingo

Compound	Dmitri Mendeleev	Liquid	Mass	Elements
Equilibrium	Nitrogen	Bond	Oxidation	Acids
Electrical Charge	Gas		Chemistry	Saturation
Periodic Table	Polymer	Hydrogen	Reaction	Metals
Matter	Energy	pH	Bunsen Burner	Atomic Number

Chemistry Bingo: Card No. 12

Chemistry Bingo

Density	Dmitri Mendeleev	Bases	Liquid	Chemistry
Bond	Energy	Hydrogen	Elements	Chlorine
Mass	Compound		Equilibrium	Gas
Osmosis	Properties	Carbon	pH	Reaction
Periodic Table	Scientific Method	Oxidation	Electrical Charge	Mole

Chemistry Bingo: Card No. 13

Chemistry Bingo

Boyle's Law	Hydrogen	Distillation	Elements	Bunsen Burner
Atomic Number	Electrical Charge	Liquid	Acids	Matter
Density	Ion		Mixture	Catalyst
Scientific Method	Properties	pH	Carbon	Bases
Periodic Table	Saturation	Chlorine	Nitrogen	Osmosis

Chemistry Bingo

Mole	Elements	Distillation	Compound	Mass
Bases	Mixture	Absorption	Bond	Bunsen Burner
Chemistry	Electrical Charge		Atom	Matter
Periodic Table	Liquid	Energy	Properties	Density
Nitrogen	Polymer	Oxidation	Corrosion	Equilibrium

Chemistry Bingo

Carbon	Liquid	Energy	Corrosion	Reactants
Saturation	Chlorine	Gas	Metals	Ion
Density	Dmitri Mendeleev		Chemistry	Equilibrium
Vacuum	Atomic Number	Periodic Table	Molecules	Reaction
Bunsen Burner	Sodium	Oxidation	Polymer	Matter

Chemistry Bingo

Catalyst	Molecules	Helium	Liquid	Boyle's Law
Mole	Bunsen Burner	Properties	Ion	Gas
Elements	Osmosis		Sodium	Energy
Scientific Method	Nitrogen	Reaction	Distillation	Chlorine
Oxygen	Density	Compound	Mass	Dmitri Mendeleev

Chemistry Bingo

Corrosion	pH	Atomic Number	Density	Saturation
Matter	Catalyst	Oxygen	Chemistry	Bunsen Burner
Elements	Chlorine		Helium	Bond
Scientific Method	Absorption	Properties	Reaction	Mixture
Sodium	Liquid	Distillation	Molecules	Bases

Chemistry Bingo: Card No. 18

Chemistry Bingo

Chemistry	Bases	Liquid	Energy	Chemistry
Mole	Mass	Matter	Compound	Ion
Molecules	Boyle's Law		Acids	Atom
Mixture	Sodium	Oxygen	Polymer	Helium
Bond	Reactants	Nitrogen	Osmosis	Oxidation

Chemistry Bingo: Card No. 19

Chemistry Bingo

pH	Molecules	Mass	Liquid	Oxidation
Atomic Number	Equilibrium	Metals	Oxygen	Saturation
Dmitri Mendeleev	Gas		Vacuum	Absorption
Surface Tension	Solid	Solution	Polymer	Sodium
Temperature	Osmosis	Reactants	Reaction	Helium

Chemistry Bingo: Card No. 20

Chemistry Bingo

Mole	Bases	Metals	Liquid	Surface Tension
Dmitri Mendeleev	Helium	Carbon	Energy	Electrical Charge
Chlorine	Nitrogen		Molecules	Distillation
Oxygen	Compound	Sodium	Scientific Method	Osmosis
Vacuum	Reactants	Oxidation	Catalyst	Polymer

Chemistry Bingo: Card No. 21

Chemistry Bingo

Corrosion	Mixture	Helium	Bond	Density
Saturation	Mass	Atom	Energy	Acids
Atomic Number	Ion		Electrical Charge	Gas
Sodium	Scientific Method	Polymer	Absorption	Boyle's Law
Reactants	Catalyst	Molecules	Chlorine	Metals

Chemistry Bingo

Carbon	Molecules	Compound	Bond	Oxidation
Bases	pH	Nitrogen	Mole	Absorption
Mixture	Density		Solution	Electrical Charge
Chlorine	Reactants	Sodium	Catalyst	Polymer
Surface Tension	Solid	Osmosis	Oxygen	Helium

© Barbara M. Peller

Chemistry Bingo

Carbon	pH	Boyle's Law	Molecules	Energy
Helium	Oxidation	Metals	Saturation	Electrical Charge
Gas	Corrosion		Density	Chlorine
Surface Tension	Solution	Sodium	Catalyst	Dmitri Mendeleev
Temperature	Vacuum	Reactants	Mass	Solid

Chemistry Bingo: Card No. 24

Chemistry Bingo

Vacuum	Metals	Molecules	Distillation	Helium
Absorption	Scientific Method	Mole	Carbon	Acids
Dmitri Mendeleev	Energy		Solution	Sodium
Atom	Surface Tension	Solid	Reactants	Ion
Oxidation	Boyle's Law	Atomic Number	Bunsen Burner	Temperature

Chemistry Bingo: Card No. 25

Chemistry Bingo

Helium	Molecules	Mixture	Saturation	Corrosion
Oxygen	Mass	Energy	pH	Carbon
Scientific Method	Solution		Ion	Vacuum
Catalyst	Bond	Surface Tension	Reactants	Sodium
Gas	Bunsen Burner	Distillation	Solid	Temperature

Chemistry Bingo

Mixture	Atomic Number	Molecules	pH	Equilibrium
Surface Tension	Solution	Mole	Sodium	Acids
Properties	Solid		Reactants	Vacuum
Corrosion	Bases	Metals	Temperature	Absorption
Bunsen Burner	Ion	Helium	Atom	Gas

Chemistry Bingo

Chemistry	pH	Atom	Molecules	Carbon
Equilibrium	Helium	Solution	Saturation	Ion
Solid	Chlorine		Gas	Oxygen
Reaction	Corrosion	Nitrogen	Reactants	Sodium
Bond	Elements	Bunsen Burner	Temperature	Surface Tension

Chemistry Bingo: Card No. 28

Chemistry Bingo

Helium	pH	Corrosion	Mole	Elements
Scientific Method	Oxygen	Metals	Gas	Atom
Dmitri Mendeleev	Solution		Acids	Molecules
Distillation	Surface Tension	Hydrogen	Reactants	Sodium
Carbon	Energy	Temperature	Bases	Solid

Chemistry Bingo

Boyle's Law	Molecules	Saturation	Elements	Sodium
Absorption	pH	Mixture	Ion	Acids
Scientific Method	Density		Gas	Metals
Temperature	Bases	Bond	Reactants	Solution
Surface Tension	Compound	Solid	Helium	Atom

Chemistry Bingo: Card No. 30

www.ingramcontent.com/pod-product-compliance
Lightning Source LLC
Chambersburg PA
CBHW051428200326
41520CB00023B/7393